Copyright © 2021 Spunky Science

All rights reserved. No part of this book may be altered, reproduced, redistributed, or used in any manner other than its original intent without written permission or copyright owner except for the use of quotation in a book review.

LAB RULES

Wash your hands

Use goggles

Do not eat or drink

Be responsible

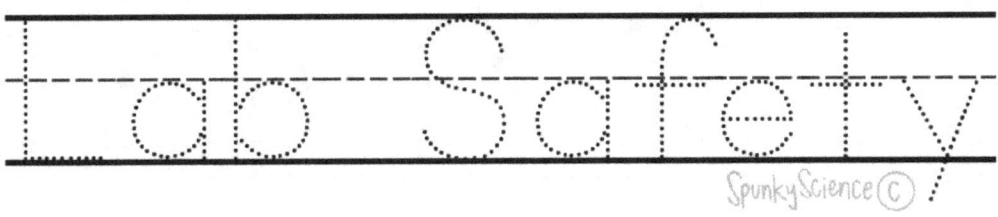

RECYCLING

Recycling is how we take trash and make it into something new. It is good for the environment because it reduces the amount of trash in the landfill.

Don't forget to recycle!

TIMER

Minutes

Seconds

A timer is used to record how long something is happening. This timer measures time in minutes and seconds and has a pause button.

Time T_____

RAIN GAUGE

A rain gauge measures how much rain fell.

RAIN TRACKER

Scientists collect data by using tools. Track the rain for the week by coloring in the amount of rain for each day. When the week is over, add them up to find the total rain for that week!

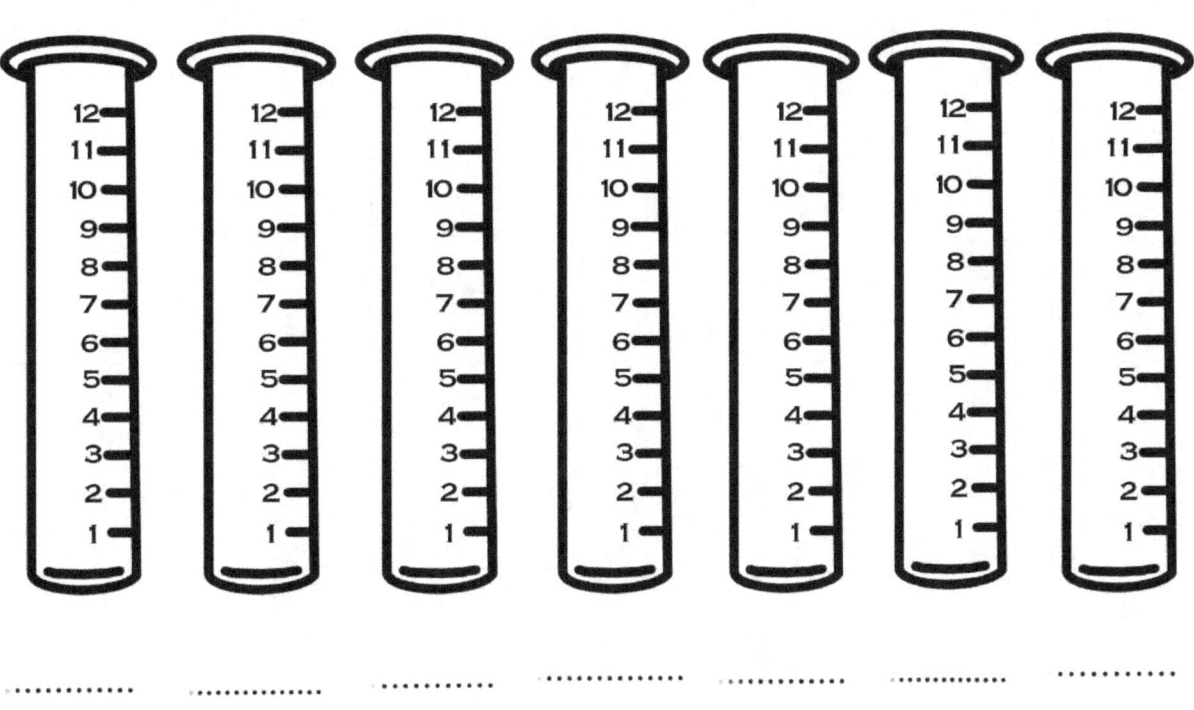

The total rain for the week is

THERMOMETER

A thermometer measures how much heat something has. If it has little heat, the number is lower and if it has a lot of heat energy, then the number is higher.

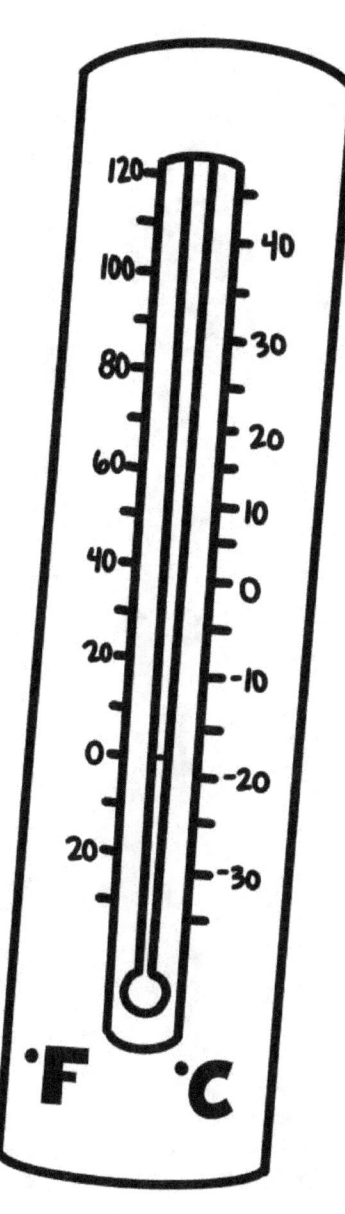

Some tell you the number while others you have to figure out.

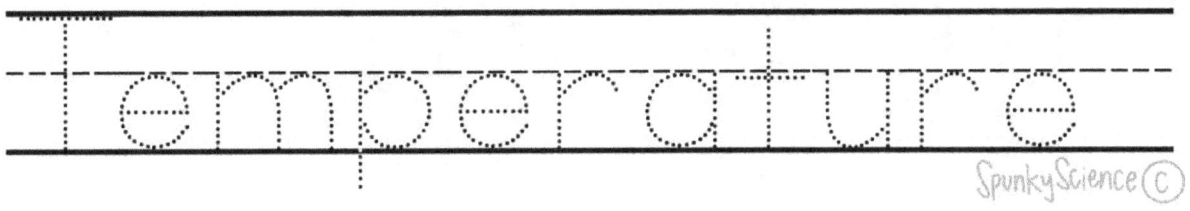

RULER

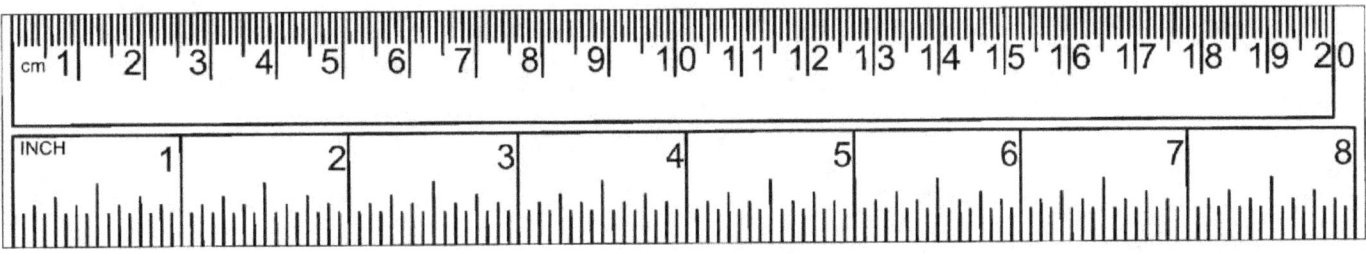

A ruler is a tool that is used to measure the length of an object. Each side measures in a different unit.

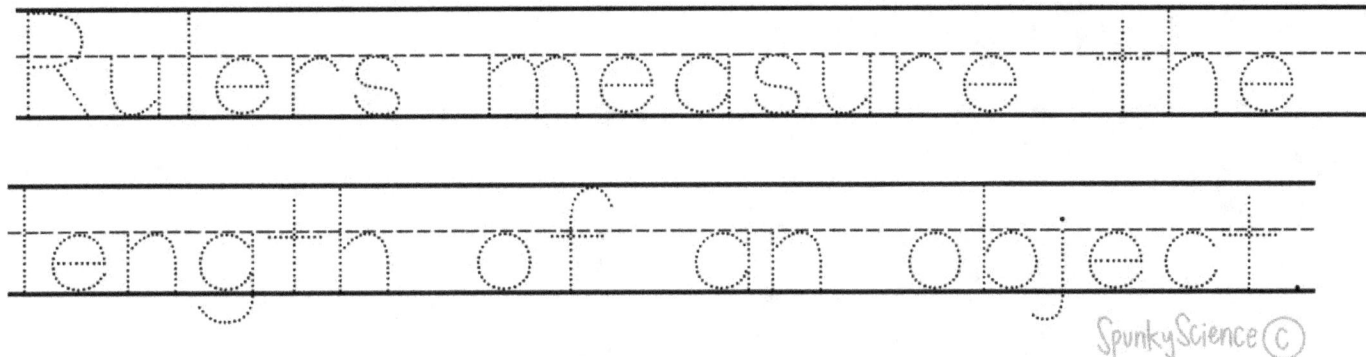

WHAT IS A TERRARIUM?

Terrarium

WHO LIVES IN AN AQUARIUM?

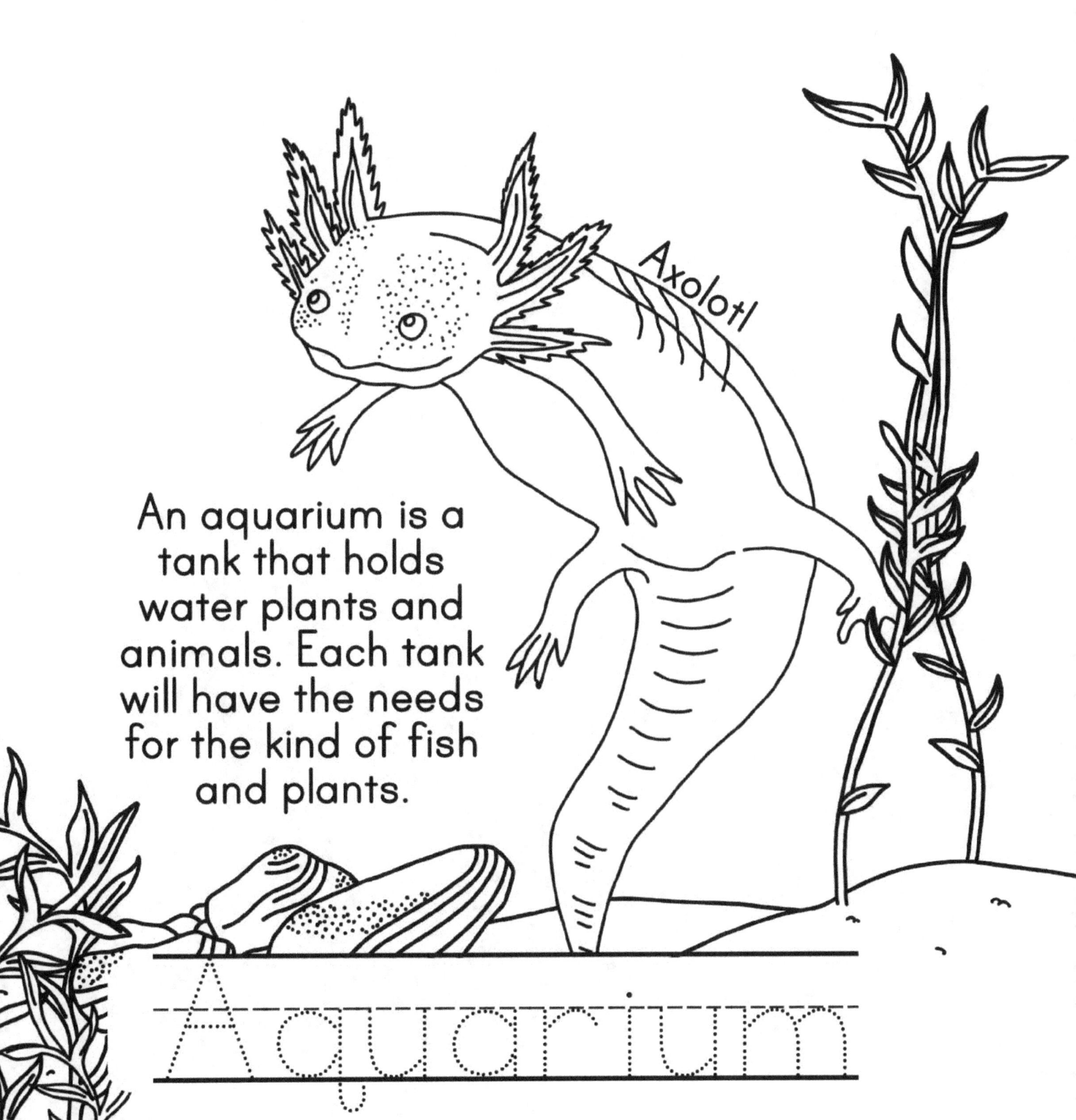

Axolotl

An aquarium is a tank that holds water plants and animals. Each tank will have the needs for the kind of fish and plants.

Aquarium

SOLID OR LIQUID?

changing states

MELTING

M _____

FREEZING

F _____

HEATING

H _____

COOLING

C _____

LIGHT

Light energy is the kind of energy that you can see. **Light travels in a straight** path.

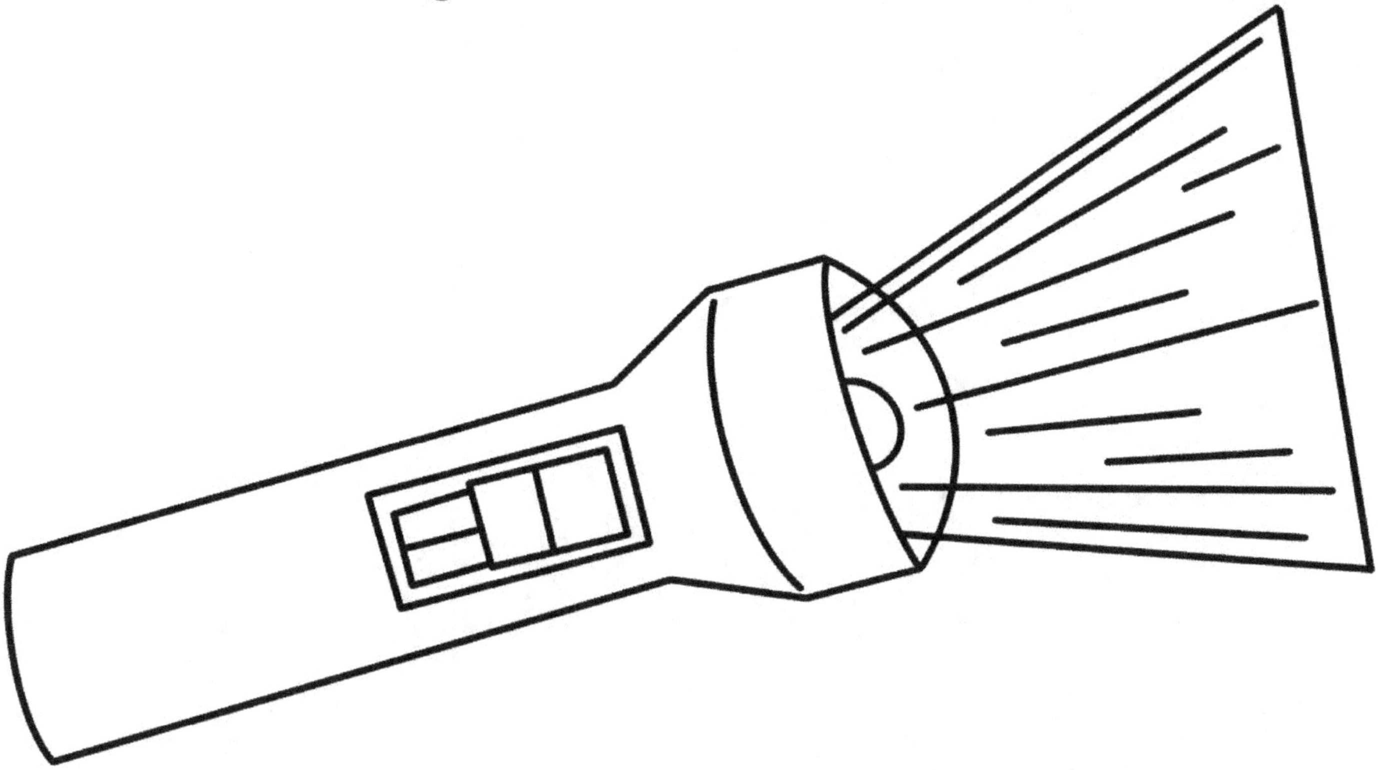

Light

WHAT IS SOUND?

Sound is caused by vibrations of an object. **In a guitar the strings vibrate which causes a** sound.

Vibrations

HOW DOES BUTTER MELT?

Heat energy is a kind of energy that you can feel. Heat moves into objects. The amount of heat can make things melt.

Heat energy

MAGNETS

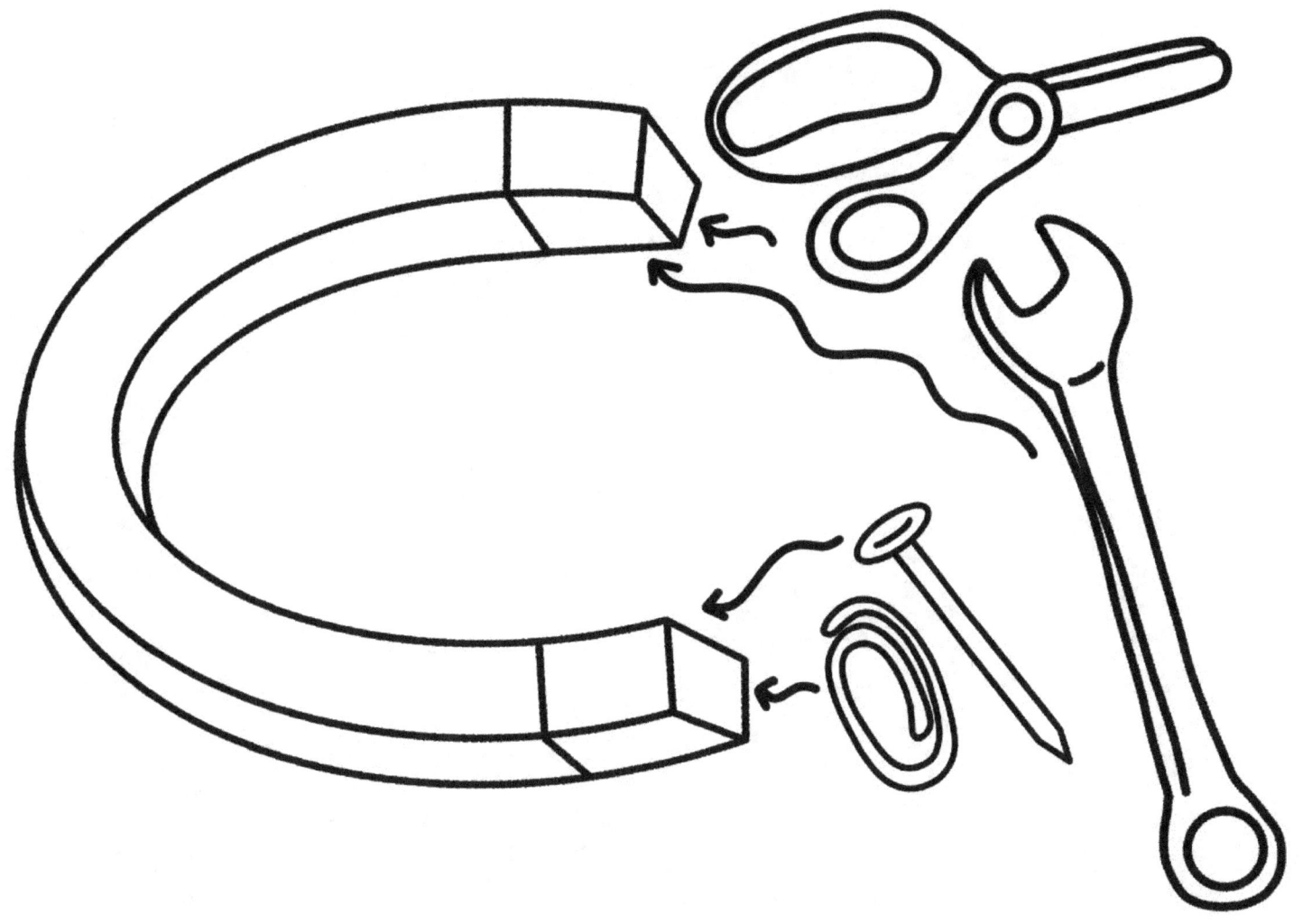

Metal objects are magnetic.

WHAT IS SAND?

ROCKS ARE CONSTANTLY BEING BROKEN DOWN BY WEATHER AND WATER. SAND IS FINELY BROKEN DOWN ROCKS.

SIZE   TEXTURE   COLOR

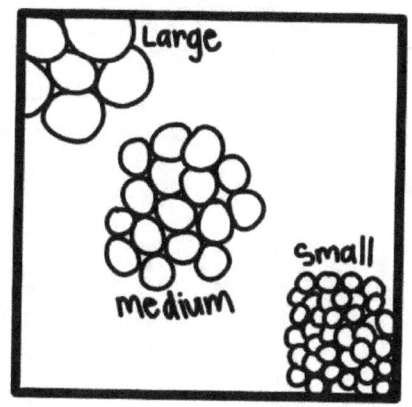 

THE COLOR DEPENDS ON THE KIND OF MINERAL THAT IT'S MADE OF.

FRESHWATER

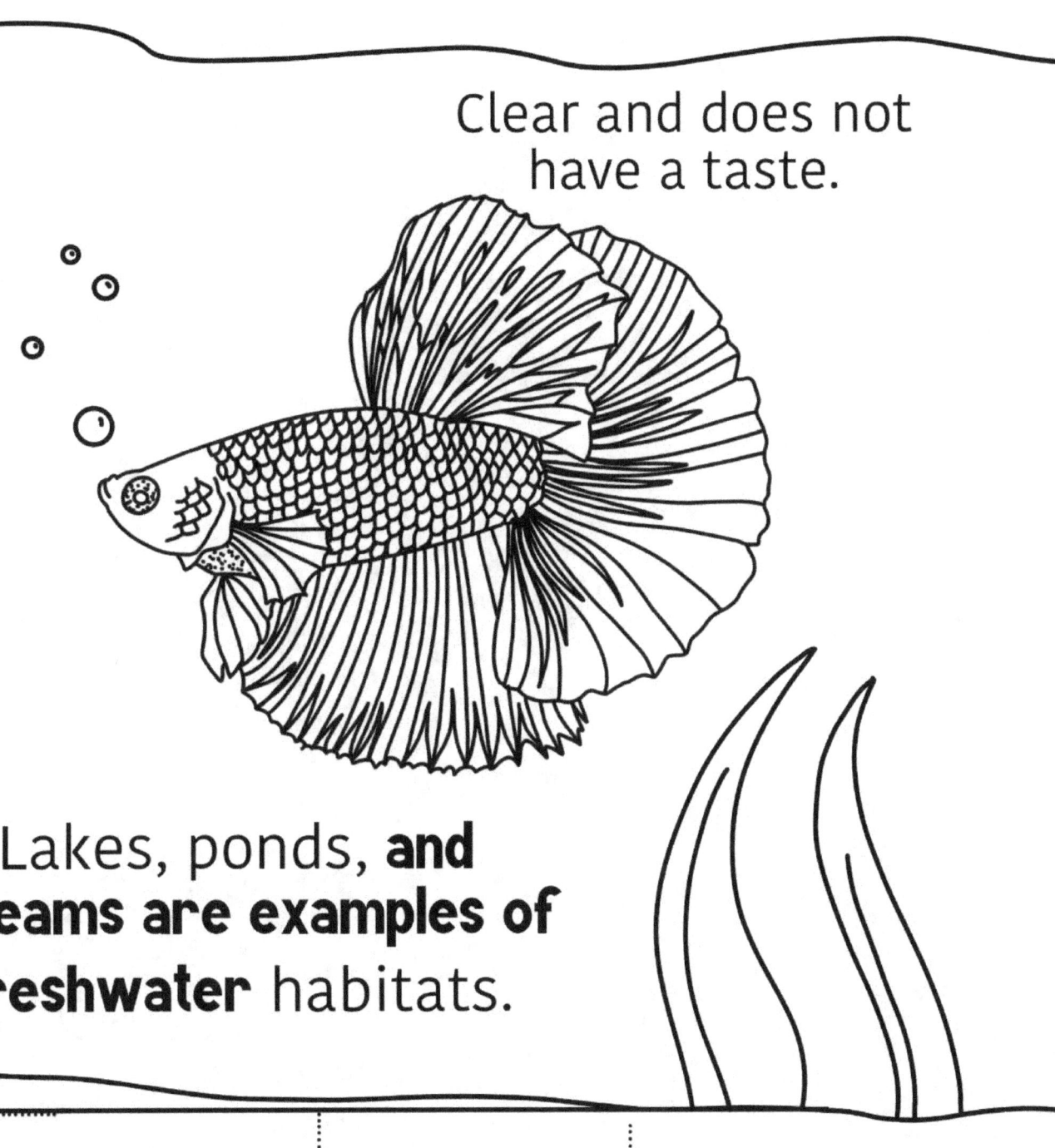

Clear and does not have a taste.

Lakes, ponds, **and streams are examples of freshwater** habitats.

Freshwater

SALTWATER

Oceans and seas are are made of saltwater. **Saltwater has more salt than** freshwater, **so that's why we can taste it.**

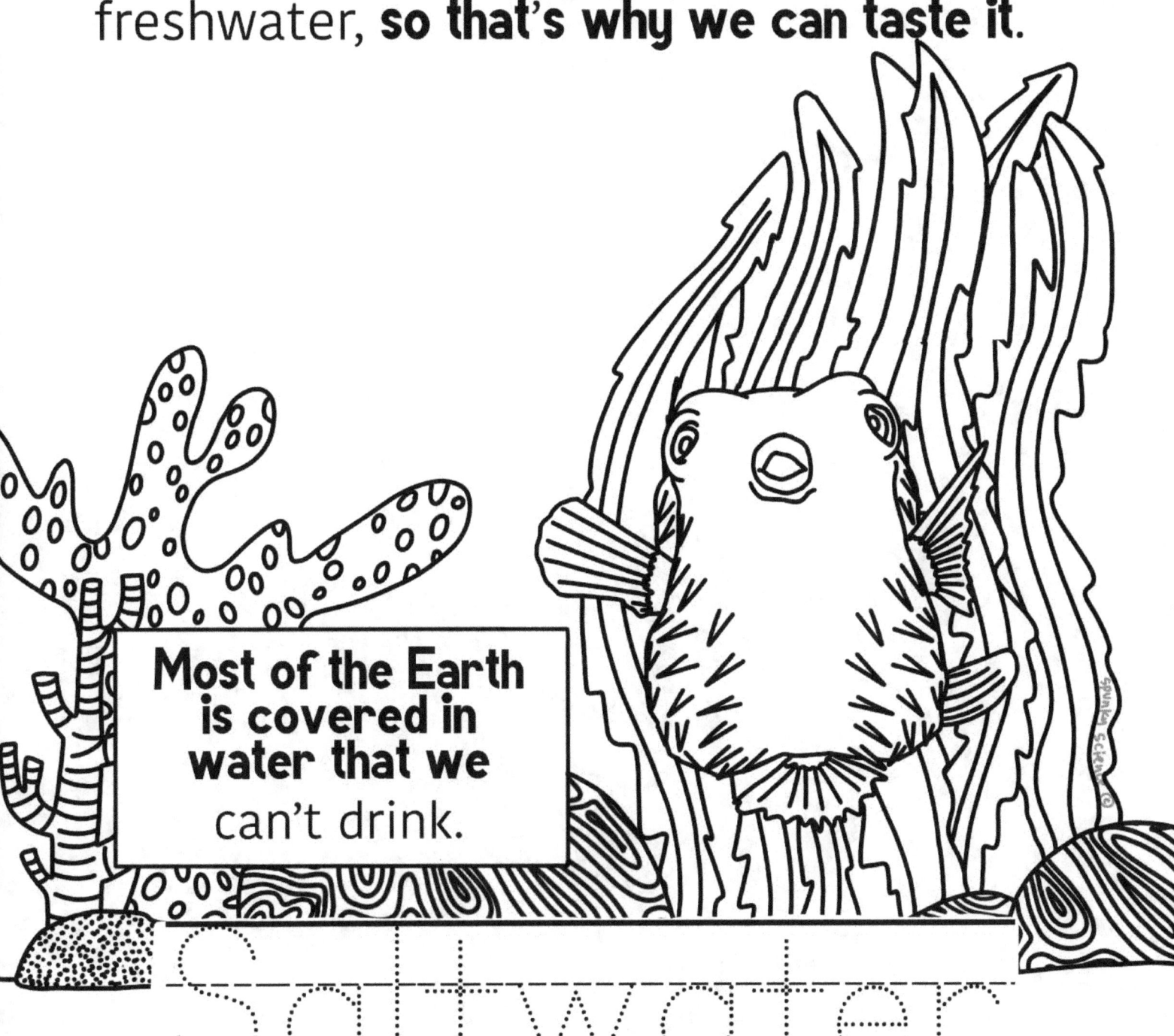

Most of the Earth is covered in water that we can't drink.

Saltwater

NATURAL RESOURCES

Natural resources are resources that are found on Earth and not made by people.

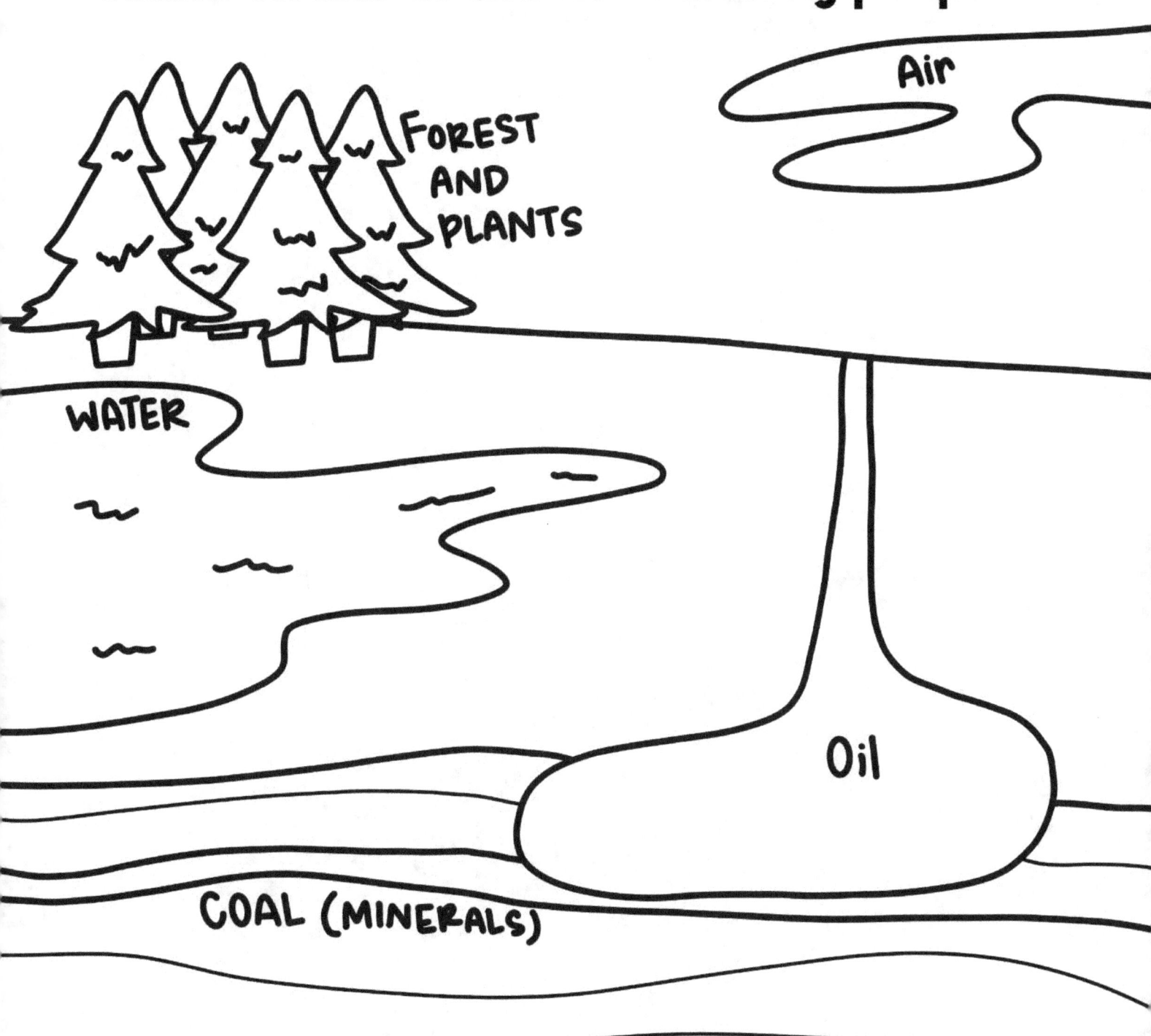

MAN MADE RESOURCES

Man made resources are things that we use that are created by people.

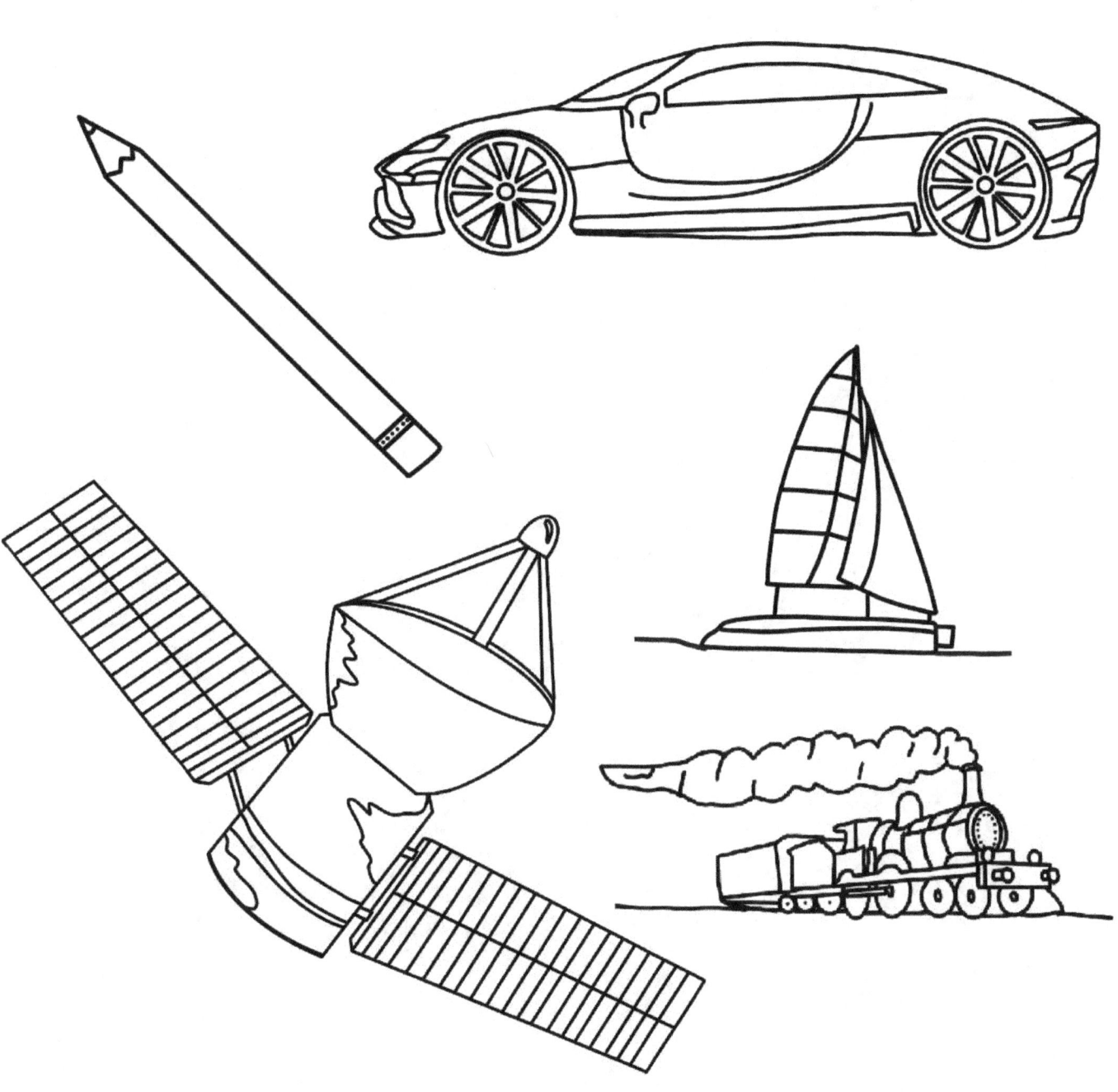

Tornado

A tornado is a funnel or fast moving air

Forms when hot air meets cold air.

tornado

Thunderstorms

Since light travels faster than sound, you see the lighting before you hear the thunder.

Thunderstorms

SUMMER

Summer is the hottest season of the year! **In the North part of the world, Summer happens between June and September. In the South part of the world, Summer takes place between December and March.**

Summer   Hot
North   South

SPRING

During Spring, **the days become longer and weather gets** warmer. **It also rains more which helps plants grow and flowers** bloom.

Spring showers bring May flowers.

FALL

Also called Autumn, **Fall is the season when weather starts getting** cooler, **leaves start changing** colors, **and animals start gathering food before Winter** arrives.

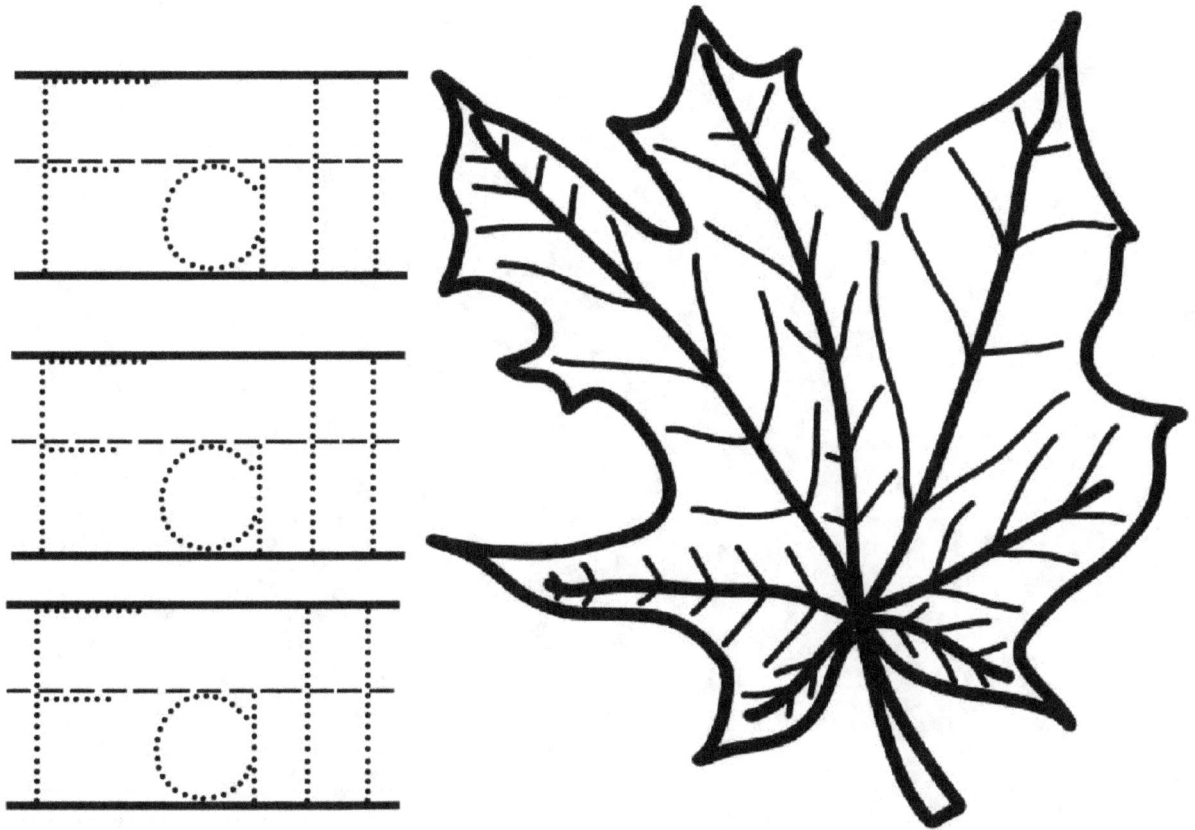

WINTER

Winter is the coldest part of the year and has the fewest hours of Sunlight. **Depending on where you** live, **it often brings** snow, ice, **and freezing** temperatures.

Ice

Snow

Winter

THE SKY

The Moon goes through different phases so it looks different depending on the day

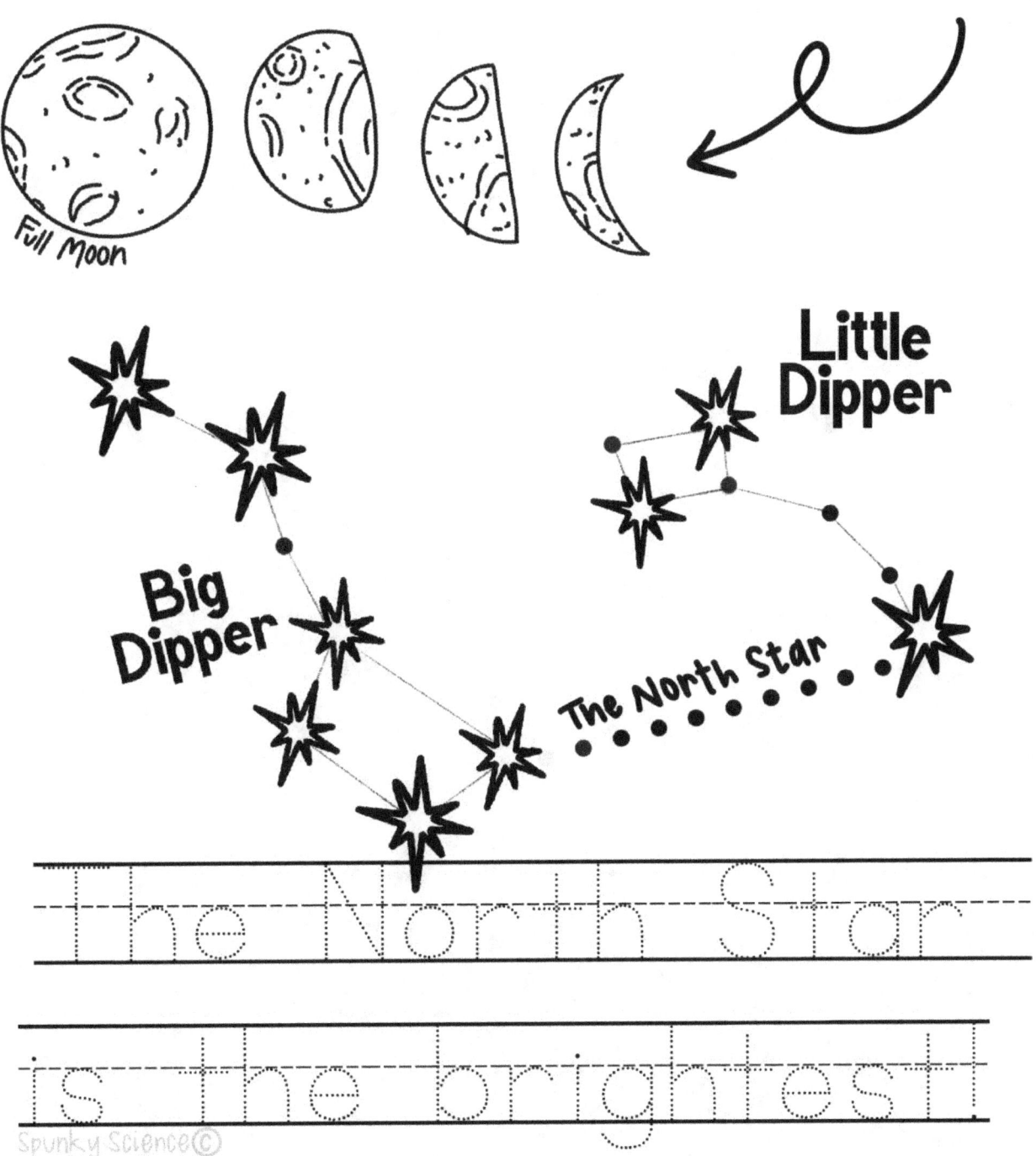

The North Star is the brightest

BASIC NEEDS OF PLANTS

SUNLIGHT

TEMPERATURE AIR

NUTRIENTS

WATER

Plants needs 5 things to survive.

WHY DO PLANTS HAVE STEMS?

- FLOWER
- BUD
- LEAF
- STEM
- ROOTS
- WATER FROM THE SOIL

Stems carry water to the rest of the plant.

Why do whales migrate?

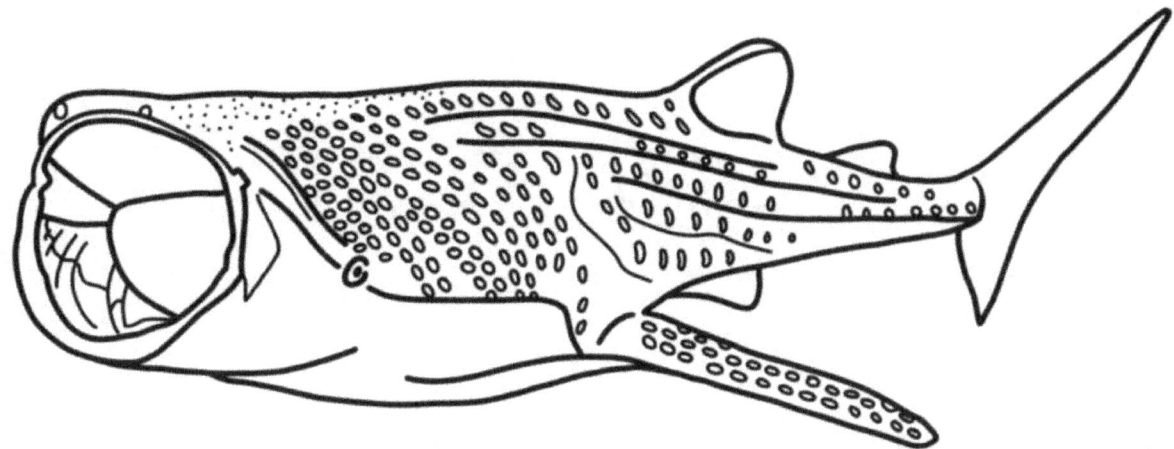

During the warm months of the year whales migrate to cold waters where the food is, then, when the weather becomes colder and the food scarce, whales will migrate to warmer water.

Migration

It may surprise you, but bats hibernate in caves during the months of October **to** April.

Hibernation

DORMANT

The pill bug enters a stage of dormancy during the winter in order to survive.

Pill bugs stay
dormant in the
Winter to stay warm.

FOOD CHAIN

Energy is always moving. **In this food** chain, **the energy moves from the** sun, **to the** plant, **and then into the** Quail.

WHY DO WOLVES HOWL?

Wolves howl for many reasons, but mostly to tell other wolves where they are located.

Wolves howl to talk to other wolves.

The lifecycle of a
BUTTERFLY

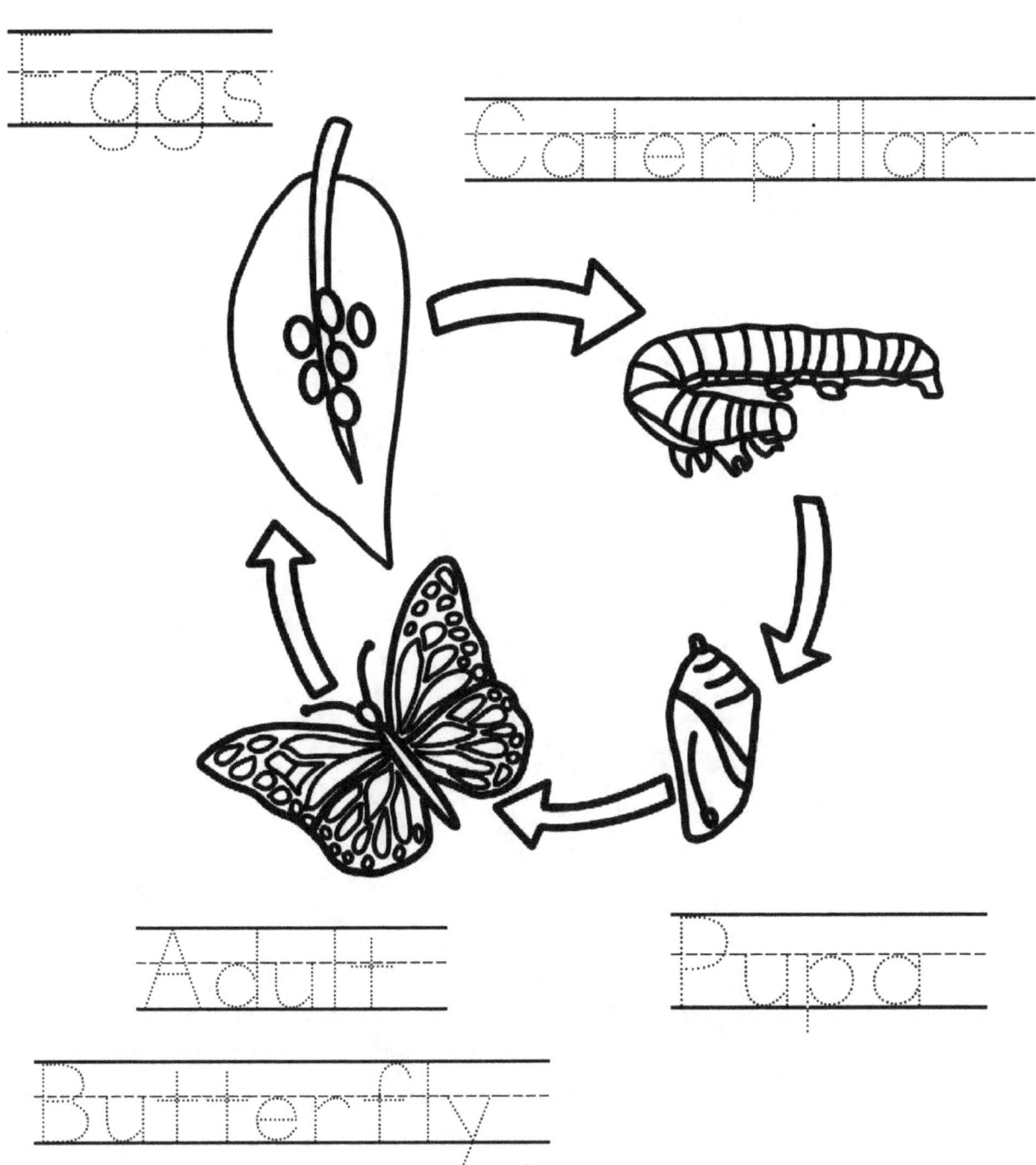

Eggs

Caterpillar

Pupa

Adult Butterfly

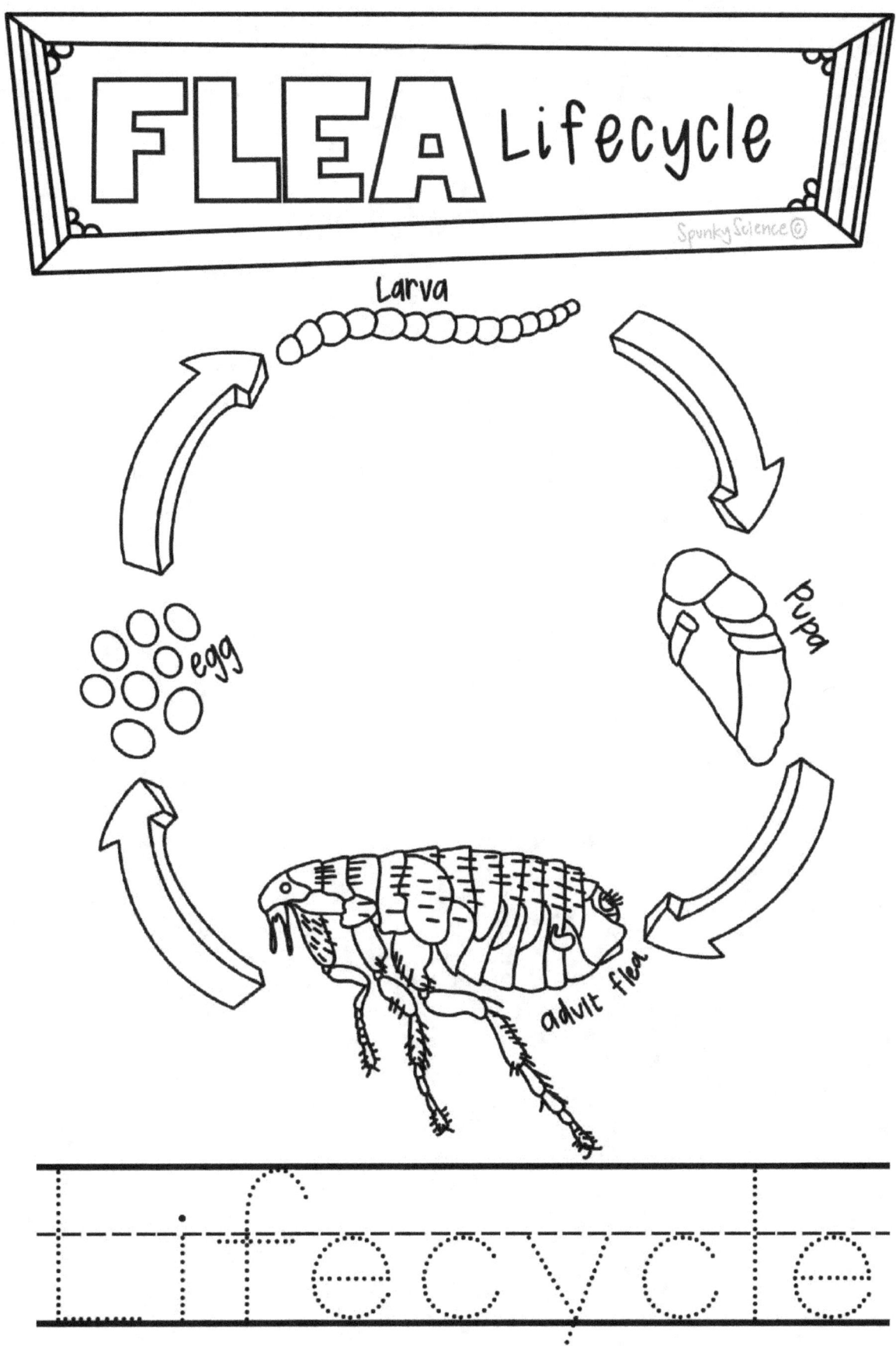

The lifecycle of a CRAB

A female crab can lay 24 million eggs in 2 years.

www.ingramcontent.com/pod-product-compliance
Lightning Source LLC
Chambersburg PA
CBHW060434220526
45465CB00008B/3142